To wonder, question, explore
— is true science.
— Coco Willi

This is the story of how the microwave was invented.

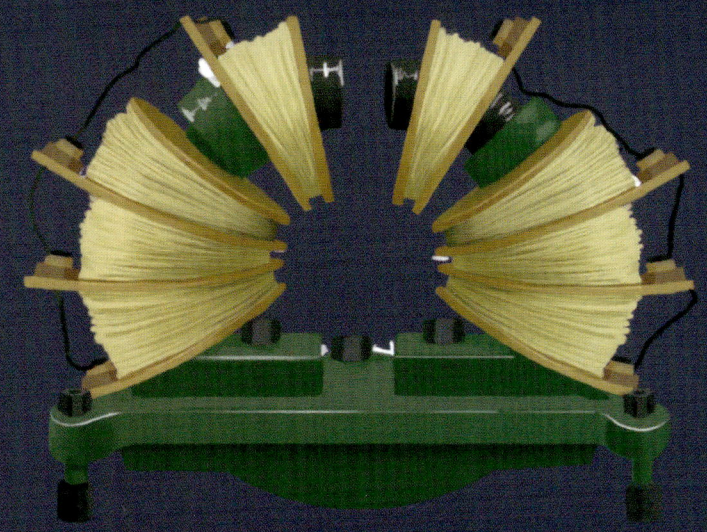

To the children
that dream, wonder,
tinker, take-apart
and build.
Never stop.

ISBN 978-1-54394-088-6
Library of Congress Catalog Number: 2018902529
Williams, Cole W.
Dr. Brainchild & Radar: A Popcorn Discovery
by Cole W. Williams; illustrations by Laura Acosta.
p. cm.

First Printing

Edition and Proofreading by Nia Howard
Design and Illustrations by Laura Acosta

Burning Belly Press
St. Paul, Minnesota
To order, visit www.colewwilliams.com.
Reseller discounts available.

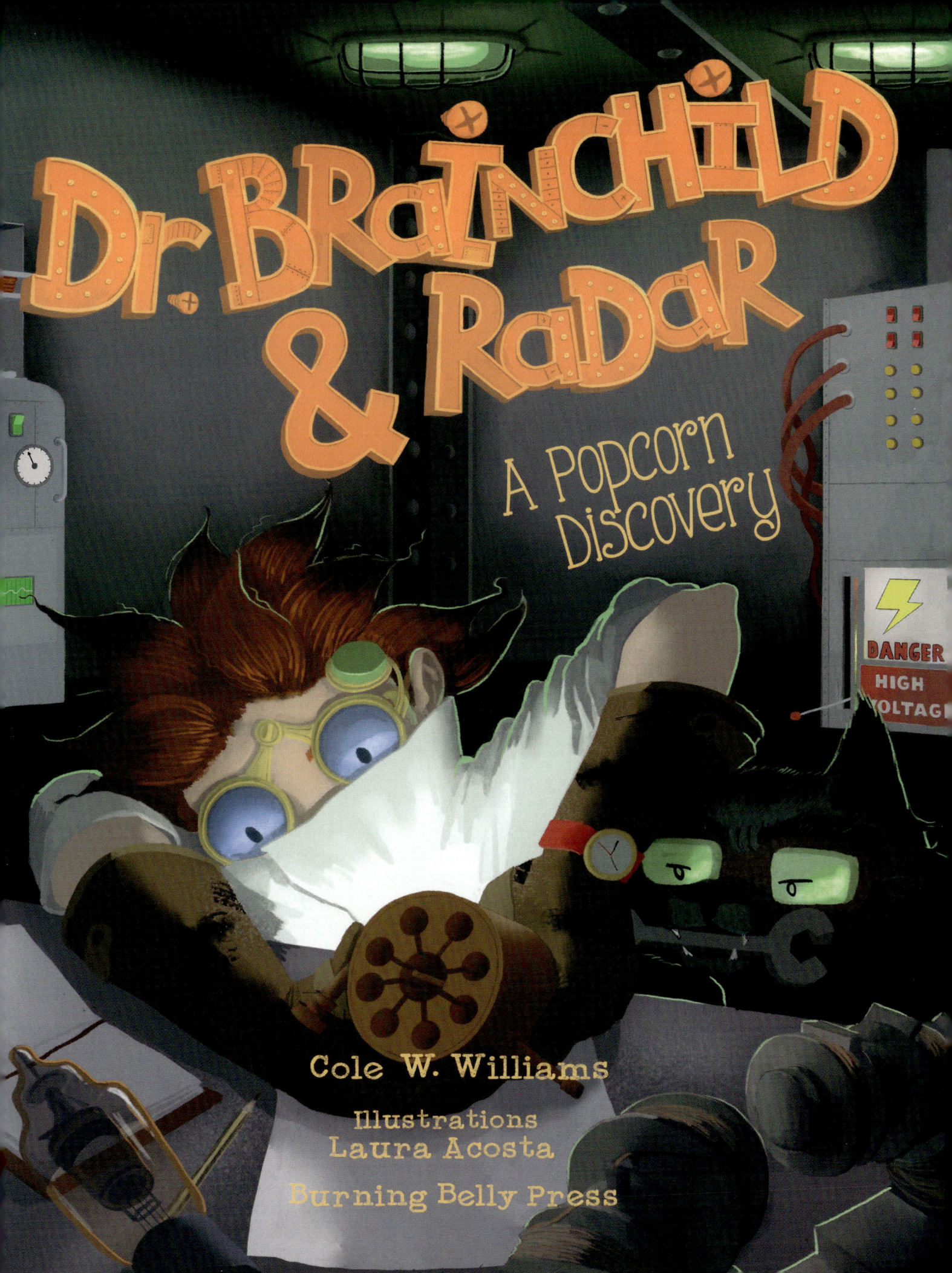

Every morning, Dr. Brainchild would get ready for work the exact same way: securing the goggles, straightening the lab coat and saying, "Today is a top-notch day! Rise and shine, Radar!"

Radar the wolf was Dr. Brainchild's assistant. Every morning, Radar would dutifully follow the wacky Doctor to the kitchen, and every morning, the Doctor would pack Radar's lunch pail with the same snack—one lonely egg.

Radar's lunch

"Save the hen fruit for lunch, Radar!
We've got work to do!"
Radar often wished for a surprise snack.

One morning, Dr. Brainchild froze on the way to the door, "Oh horse feathers! I almost forgot..."
Dr. Brainchild reached for the chocolate bar on the side table and proclaimed, "Helps me think, Radar!"

CHOCOLATE

And without a thought, Dr. Brainchild tucked the chocolate bar into the front pocket of the lab coat.

Down one flight of stairs, a turn to the right and exactly twenty-two extra steps, Dr. Brainchild and Radar reached the giant metal door to their laboratory.

The duo's goal was to perfect radio waves in the air. On test day, Dr. Brainchild asked Radar to turn up the power surging through the Magnetron.
"Crank up the volts, Radar! Ha-hee!"

As the Magnetron hummed like a giant swarm of insects, Dr. Brainchild began to smell the rich aroma of chocolate and looked down, only to notice the sweet treat was melting inside the lab coat pocket!

"Applesauce, Radar, my chocolate! How in the Milky Way did that happen?"

As Dr. Brainchild devised a new experiment, Radar slept on the floor and had a lovely dream about a river of chocolate flowing through the lab.

The next morning began with a bang!
"Radar! Hen fruits! We are going to need them!"
And before Radar even made it to the kitchen, Dr. Brainchild was off to the lab in a whirlwind
of frenzied excitement.

"Be Bop!" Radar, put a hop in your
step!" Dr. Brainchild hollered.

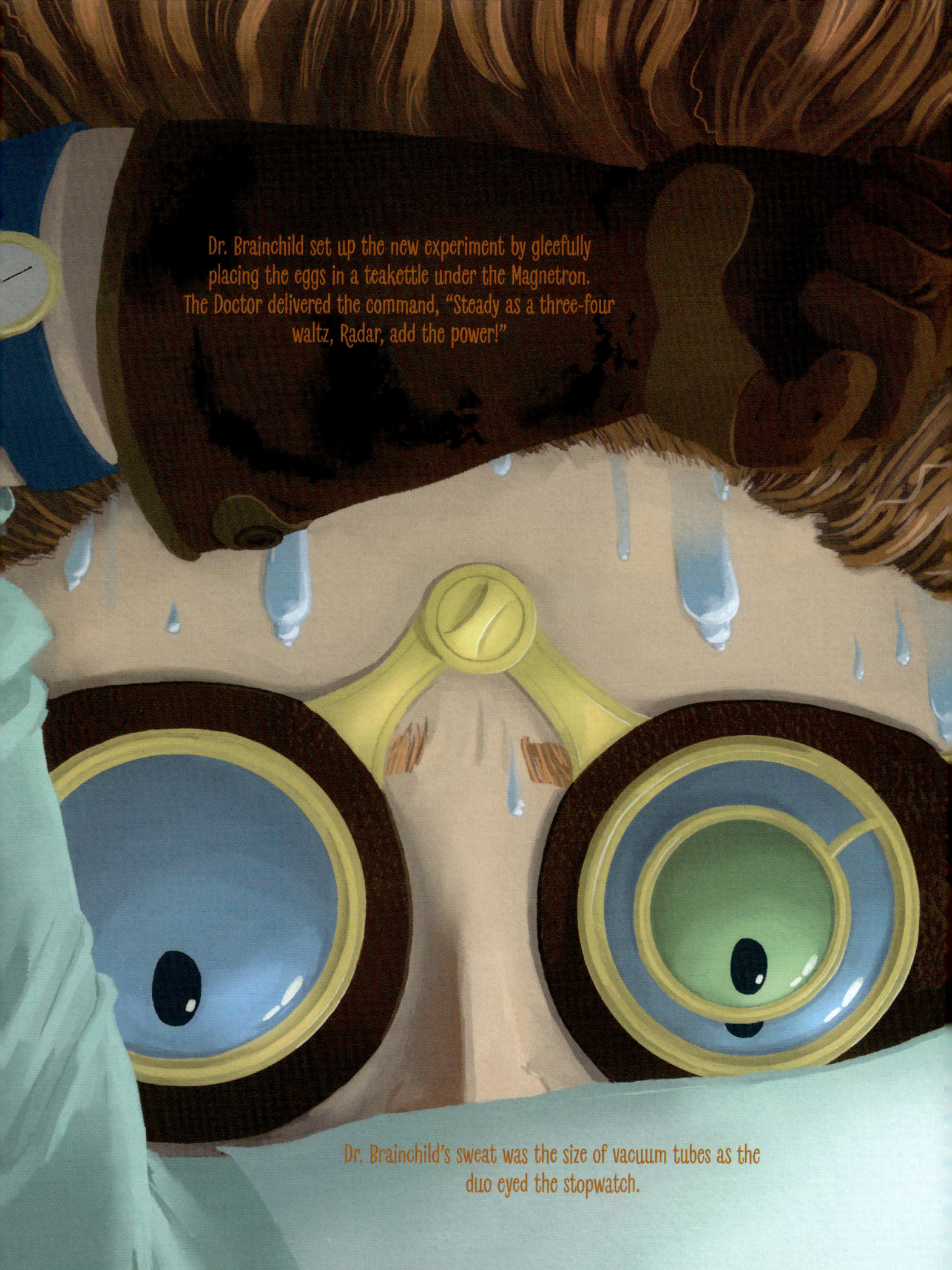

Dr. Brainchild set up the new experiment by gleefully placing the eggs in a teakettle under the Magnetron. The Doctor delivered the command, "Steady as a three-four waltz, Radar, add the power!"

Dr. Brainchild's sweat was the size of vacuum tubes as the duo eyed the stopwatch.

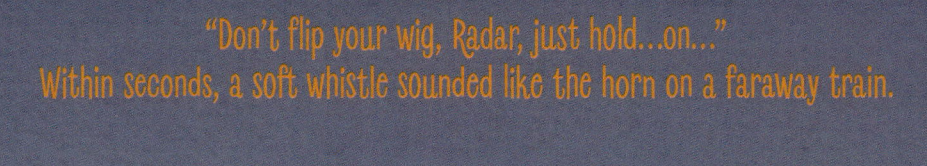

"Don't flip your wig, Radar, just hold...on..."
Within seconds, a soft whistle sounded like the horn on a faraway train.

The eggs began to sweat too, rocking back and forth,
when suddenly, a glorious "Pffft-bang!" sounded, and hot
cooked egg chunks splattered onto the Doctor's goggles.
"Scrambled eggs, whoopee!"

The metal lab door swung open, and for the second time that day, Dr. Brainchild was on the run.
"I will be back in a jiffy!"
Dr. Brainchild did cartwheels at the
thought of a new experiment and
tap danced all the way into town.

Out of breath and without wasting time, the Doctor arrived
at a popcorn cart, "How much for kernels?"

"Kernels you say?" Auntie May at Aunt May's Popcorn
Delights saw no trouble in selling just the kernels, but she
didn't understand why.
"So, you say you want this popcorn, un-popped?"

"Yes, I most terrifically would! And I thank you kindly and profusely."
Dr. Brainchild handed off the bills to pay for the seeds.

With lightning strikes under foot and kernels bouncing to the beat, Dr. Brainchild returned to the laboratory. "What a blast science is, Radar! Do we have a metal box anywhere?"

Radar slid a metal box across the lab table.
And with that, Dr. Brainchild poured Auntie May's kernels into a bowl, added oil
and placed the bowl under the metal box and the metal box under the device.

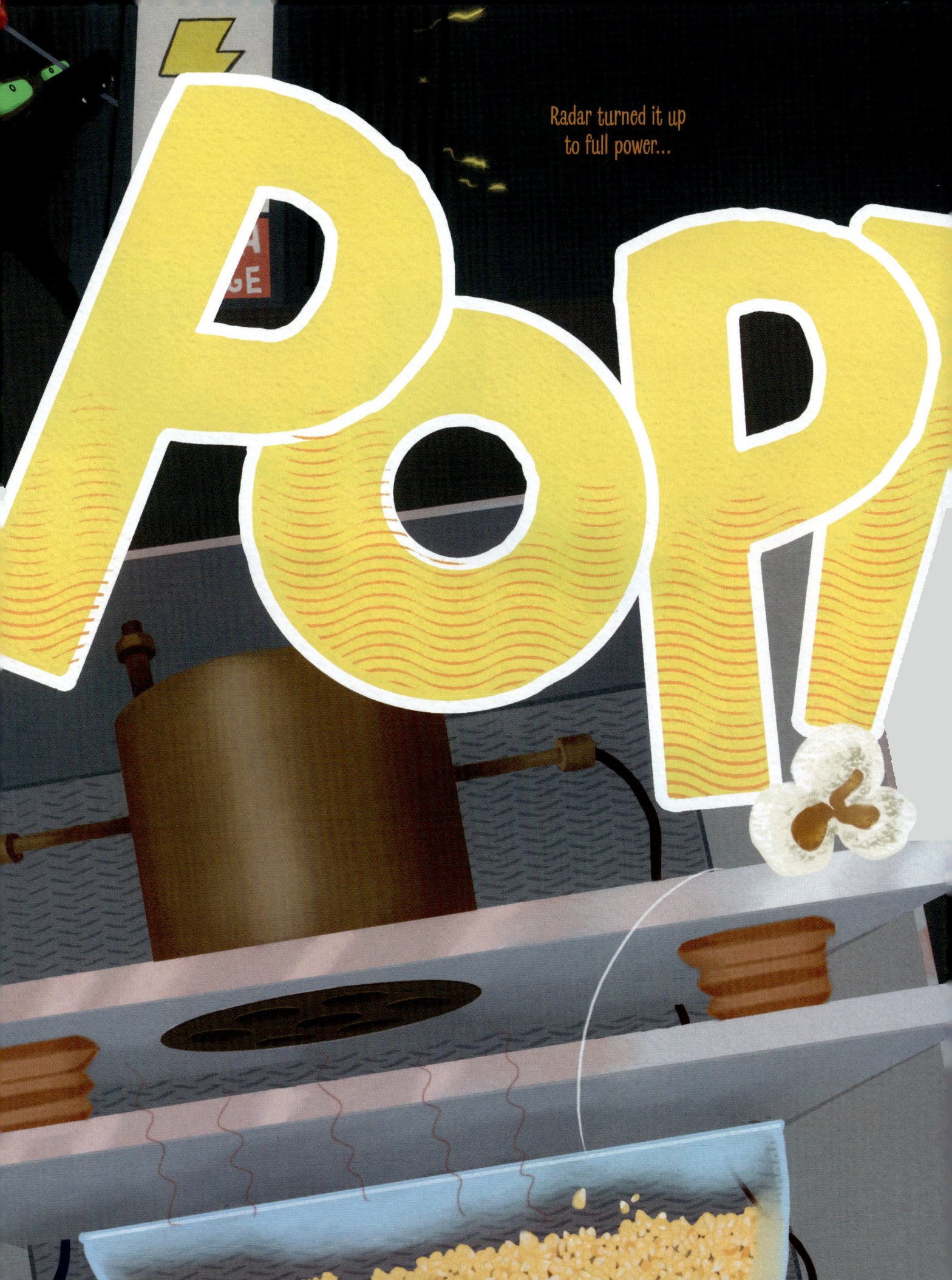

Radar turned it up
to full power...

POP

And suddenly a

PoP!

And another

PoP! And another!

Until a swarm of soft popcorn puffs whizzed across the room...

Kernels bounced off the walls like stones on a tin can
"Ting-ting, ting-ting-ting!"

Radar leapt through the air with his
jaws wide open. He trapped streams of
puffballs in his mouth; his dream had
come true.

With the power left out of control, the growing mass of popcorn began to lift the metal box off the table; a rumbling tide overtook the equipment.

From that day forward, the dynamic duo toiled over the machine, until one day, they perfected it—The Microwave—and they filled the hot box with an endless supply of popcorn.

"Radar..." Dr. Brainchild said, "I think we need to invent a popcorn bag..."

The first microwave was made in 1947. It was about six feet tall and weighed 750 pounds; that is roughly the size of a modern refrigerator—but heavier!

American engineer Percy Spencer is lauded with inventing the microwave. Dr. Brainchild's magnetron and egg experiments are fashioned after the real experiments Spencer used in his discovery!

A magnetron is a complicated device that creates energetic, short-length radio waves that travel at the speed of light. The magnetron produces microwave radiation, which agitates the water molecules in food. As these molecules become agitated (or dance), they generate heat. This heat is what cooks the food in a microwave!

To learn more about the machine that cooks your after-school snacks, hop over to www.explainthatstuff.com/how-magnetrons-work.html